Student Resources

INCLUDES
- Program Authors
- Table of Contents
- Glossary
- Common Core State Standards Correlation
- Index
- Table of Measures

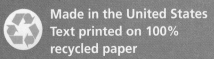

Made in the United States
Text printed on 100% recycled paper

Copyright © by Houghton Mifflin Harcourt Publishing Company

All rights reserved. No part of this work may be reproduced or transmitted in any form or by any means, electronic or mechanical, including photocopying or recording, or by any information storage or retrieval system, without the prior written permission of the copyright owner unless such copying is expressly permitted by federal copyright law.

Permission is hereby granted to individuals using the corresponding student's textbook or kit as the major vehicle for regular classroom instruction to photocopy entire pages from this publication in classroom quantities for instructional use and not for resale. Requests for information on other matters regarding duplication of this work should be addressed to Houghton Mifflin Harcourt Publishing Company, Attn: Contracts, Copyrights, and Licensing, 9400 Southpark Center Loop, Orlando, Florida 32819-8647.

Common Core State Standards © Copyright 2010. National Governors Association Center for Best Practices and Council of Chief State School Officers. All rights reserved.

This product is not sponsored or endorsed by the Common Core State Standards Initiative of the National Governors Association Center for Best Practices and the Council of Chief State School Officers.

Printed in the U.S.A.

ISBN 978-0-544-34146-3

6 7 8 9 10   0928   22 21 20 19 18 17 16 15

4500565487   B C D E F G

> If you have received these materials as examination copies free of charge, Houghton Mifflin Harcourt Publishing Company retains title to the materials and they may not be resold. Resale of examination copies is strictly prohibited.

> Possession of this publication in print format does not entitle users to convert this publication, or any portion of it, into electronic format.

Dear Students and Families,

Welcome to **Go Math!**, Grade 6! In this exciting mathematics program, there are hands-on activities to do and real-world problems to solve. Best of all, you will write your ideas and answers right in your book. In **Go Math!**, writing and drawing on the pages helps you think deeply about what you are learning, and you will really understand math!

By the way, all of the pages in your **Go Math!** book are made using recycled paper. We wanted you to know that you can Go Green with **Go Math!**

Sincerely,

The Authors

Made in the United States
Text printed on 100% recycled paper

GO MATH!

Authors

Juli K. Dixon, Ph.D.
Professor, Mathematics Education
University of Central Florida
Orlando, Florida

Edward B. Burger, Ph.D.
President, Southwestern University
Georgetown, Texas

Steven J. Leinwand
Principal Research Analyst
American Institutes for
   Research (AIR)
Washington, D.C.

Contributor

Rena Petrello
Professor, Mathematics
Moorpark College
Moorpark, CA

Matthew R. Larson, Ph.D.
K-12 Curriculum Specialist for
   Mathematics
Lincoln Public Schools
Lincoln, Nebraska

Martha E. Sandoval-Martinez
Math Instructor
El Camino College
Torrance, California

English Language Learners Consultant

Elizabeth Jiménez
CEO, GEMAS Consulting
Professional Expert on English
   Learner Education
Bilingual Education and
   Dual Language
Pomona, California

Table of Contents

Student Edition Table of Contents.......................v
Glossary...H1
Common Core State Standards Correlation..............H17
Index...H27
Table of Measures.................................H39

The Number System

 Critical Area Completing understanding of division of fractions and extending the notion of number to the system of rational numbers, which includes negative numbers

Real World Project   Sweet Success 2

1 Whole Numbers and Decimals 3

Domain The Number System
COMMON CORE STATE STANDARDS 6.NS.B.2, 6.NS.B.3, 6.NS.B.4

- ✓ Show What You Know 3
- Vocabulary Builder 4
- 1 Divide Multi-Digit Numbers 5
- 2 Prime Factorization 11
- 3 Least Common Multiple 17
- 4 Greatest Common Factor 23
- 5 **Problem Solving** • Apply the Greatest Common Factor 29
- ✓ **Mid-Chapter Checkpoint** 35
- 6 Add and Subtract Decimals 37
- 7 Multiply Decimals 43
- 8 Divide Decimals by Whole Numbers 49
- 9 Divide with Decimals 55
- ✓ **Chapter 1 Review/Test** 61

2 Fractions 67

Domain The Number System
COMMON CORE STATE STANDARDS 6.NS.A.1, 6.NS.B.4, 6.NS.C.6c

- ✓ Show What You Know 67
- Vocabulary Builder 68
- 1 Fractions and Decimals 69
- 2 Compare and Order Fractions and Decimals 75
- 3 Multiply Fractions 81
- 4 Simplify Factors 87
- ✓ **Mid-Chapter Checkpoint** 93
- 5 Investigate • Model Fraction Division 95
- 6 Estimate Quotients 101
- 7 Divide Fractions 107
- 8 Investigate • Model Mixed Number Division 113
- 9 Divide Mixed Numbers 119
- 10 **Problem Solving** • Fraction Operations 125
- ✓ **Chapter 2 Review/Test** 131

Critical Area

GO DIGITAL

Go online! Your math lessons are interactive. Use iTools, Animated Math Models, the Multimedia eGlossary, and more.

Chapter 1 Overview
In this chapter, you will explore and discover answers to the following **Essential Questions**:
- How do you solve real-world problems involving whole numbers and decimals?
- How does estimation help you solve problems involving decimals and whole numbers?
- How can you use the GCF and the LCM to solve problems?

Chapter 2 Overview
In this chapter, you will explore and discover answers to the following **Essential Questions**:
- How can you use the relationship between multiplication and division to divide fractions?
- What is a mixed number?
- How can you estimate products and quotients of fractions and mixed numbers?

v

Chapter 3 Overview

In this chapter, you will explore and discover answers to the following **Essential Questions**:

- How do you write, interpret, and use rational numbers?
- How do you calculate the absolute value of a number?
- How do you graph an ordered pair?

Practice and Homework

Lesson Check and Spiral Review in every lesson

3 Rational Numbers — 137

Domain The Number System
COMMON CORE STATE STANDARDS 6.NS.C.5, 6.NS.C.6a, 6.NS.C.6b, 6.NS.C.6c, 6.NS.C.7a, 6.NS.C.7b, 6.NS.C.7c, 6.NS.C.7d, 6.NS.C.8

✓ Show What You Know 137
　Vocabulary Builder 138
1　Understand Positive and Negative Numbers 139
2　Compare and Order Integers 145
3　Rational Numbers and the Number Line 151
4　Compare and Order Rational Numbers 157
✓ Mid-Chapter Checkpoint 163
5　Absolute Value 165
6　Compare Absolute Values 171
7　Rational Numbers and the Coordinate Plane 177
8　Ordered Pair Relationships 183
9　Distance on the Coordinate Plane 189
10　Problem Solving • The Coordinate Plane 195
✓ Chapter 3 Review/Test 201

Ratios and Rates

 Critical Area Connecting ratio and rate to whole number multiplication and division and using concepts of ratio and rate to solve problems

Real World Project Meet Me in St. Louis............................208

4 Ratios and Rates   209

Domain Ratios and Proportional Relationships
COMMON CORE STATE STANDARDS 6.RP.A.1, 6.RP.A.2, 6.RP.A.3a, 6.RP.A.3b

- ✓ Show What You Know..209
- Vocabulary Builder..210
- 1 Investigate • Model Ratios.................................211
- 2 Ratios and Rates..217
- 3 Equivalent Ratios and Multiplication Tables................223
- 4 Problem Solving • Use Tables to Compare Ratios............229
- 5 Algebra • Use Equivalent Ratios............................235
- ✓ Mid-Chapter Checkpoint.....................................241
- 6 Find Unit Rates...243
- 7 Algebra • Use Unit Rates...................................249
- 8 Algebra • Equivalent Ratios and Graphs.....................255
- ✓ Chapter 4 Review/Test......................................261

5 Percents   267

Domain Ratios and Proportional Relationships
COMMON CORE STATE STANDARDS 6.RP.A.3c

- ✓ Show What You Know..267
- Vocabulary Builder..268
- 1 Investigate • Model Percents...............................269
- 2 Write Percents as Fractions and Decimals...................275
- 3 Write Fractions and Decimals as Percents...................281
- ✓ Mid-Chapter Checkpoint.....................................287
- 4 Percent of a Quantity......................................289
- 5 Problem Solving • Percents.................................295
- 6 Find the Whole from a Percent..............................301
- ✓ Chapter 5 Review/Test......................................307

Critical Area

GO DIGITAL
Go online! Your math lessons are interactive. Use iTools, Animated Math Models, the Multimedia eGlossary, and more.

Chapter 4 Overview
In this chapter, you will explore and discover answers to the following **Essential Questions**:
- How can you use ratios to express relationships and solve problems?
- How can you write a ratio?
- What are equivalent ratios?
- How are rates related to ratios?

Chapter 5 Overview
In this chapter, you will explore and discover answers to the following **Essential Questions**:
- How can you use ratio reasoning to solve percent problems?
- How can you write a percent as a fraction?
- How can you use a ratio to find a percent of a number?

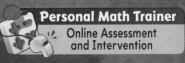
Personal Math Trainer
Online Assessment and Intervention

Chapter 6 Overview

In this chapter, you will explore and discover answers to the following **Essential Questions**:

- How can you use measurements to help you describe and compare objects?
- Why do you need to convert between units of measure?
- How can you use a ratio to convert units?
- How do you transform units to solve problems?

Practice and Homework

Lesson Check and Spiral Review in every lesson

6 Units of Measure 313

Domain Ratios and Proportional Relationships
COMMON CORE STATE STANDARDS 6.RP.A.3d

- ✓ Show What You Know 313
- Vocabulary Builder 314
- **1** Convert Units of Length 315
- **2** Convert Units of Capacity 321
- **3** Convert Units of Weight and Mass 327
- ✓ Mid-Chapter Checkpoint 333
- **4** Transform Units 335
- **5** Problem Solving • Distance, Rate, and Time Formulas 341
- ✓ Chapter 6 Review/Test 347

Expressions and Equations

 Critical Area Writing, interpreting, and using expressions and equations

Real World Project The Great Outdoors 354

7 Algebra: Expressions 355

Domain Expressions and Equations
COMMON CORE STATE STANDARDS 6.EE.A.1, 6.EE.A.2a, 6.EE.A.2b, 6.EE.A.2c, 6.EE.A.3, 6.EE.A.4, 6.EE.B.6

✓ Show What You Know 355
   Vocabulary Builder 356
1 Exponents 357
2 Evaluate Expressions Involving Exponents 363
3 Write Algebraic Expressions 369
4 Identify Parts of Expressions 375
5 Evaluate Algebraic Expressions and Formulas 381
✓ Mid-Chapter Checkpoint 387
6 Use Algebraic Expressions 389
7 **Problem Solving** • Combine Like Terms 395
8 Generate Equivalent Expressions 401
9 Identify Equivalent Expressions 407
✓ Chapter 7 Review/Test 413

GO DIGITAL
Go online! Your math lessons are interactive. Use iTools, Animated Math Models, the Multimedia eGlossary, and more.

Chapter 7 Overview
In this chapter, you will explore and discover answers to the following **Essential Questions**:
- How do you write, interpret, and use algebraic expressions?
- How can you use expressions to represent real-world situations?
- How do you use the order of operations to evaluate expressions?
- How can you tell whether two expressions are equivalent?

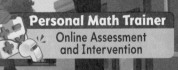
Personal Math Trainer
Online Assessment and Intervention

Chapter 8 Overview

In this chapter, you will explore and discover answers to the following **Essential Questions**:

- How can you use equations and inequalities to represent situations and solve problems?
- How can you use Properties of Equality to solve equations?
- How do inequalities differ from equations?
- Why is it useful to describe situations by using algebra?

Practice and Homework

Lesson Check and Spiral Review in every lesson

Chapter 9 Overview

In this chapter, you will explore and discover answers to the following **Essential Questions**:

- How can you show relationships between variables?
- How can you determine the equation that gives the relationship between two variables?
- How can you use tables and graphs to visualize the relationship between two variables?

---

8 Algebra: Equations and Inequalities — 419

Domain Expressions and Equations
COMMON CORE STATE STANDARDS 6.EE.B.5, 6.EE.B.7, 6.EE.B.8

- ✓ Show What You Know 419
- Vocabulary Builder 420
- 1 Solutions of Equations 421
- 2 Write Equations 427
- 3 Investigate • Model and Solve Addition Equations 433
- 4 Solve Addition and Subtraction Equations 439
- 5 Investigate • Model and Solve Multiplication Equations . . . 445
- 6 Solve Multiplication and Division Equations 451
- 7 Problem Solving • Equations with Fractions 457
- ✓ Mid-Chapter Checkpoint 463
- 8 Solutions of Inequalities 465
- 9 Write Inequalities 471
- 10 Graph Inequalities 477
- ✓ Chapter 8 Review/Test 483

9 Algebra: Relationships Between Variables — 489

Domain Expressions and Equations
COMMON CORE STATE STANDARDS 6.EE.C.9

- ✓ Show What You Know 489
- Vocabulary Builder 490
- 1 Independent and Dependent Variables 491
- 2 Equations and Tables 497
- 3 Problem Solving • Analyze Relationships 503
- ✓ Mid-Chapter Checkpoint 509
- 4 Graph Relationships 511
- 5 Equations and Graphs 517
- ✓ Chapter 9 Review/Test 523

Geometry and Statistics

 Critical Area Solve real-world and mathematical problems involving area, surface area, and volume; and developing understanding of statistical thinking

Real World Project This Place Is a Zoo!. **530**

10 Area 531
Domain Geometry
COMMON CORE STATE STANDARDS 6.G.A.1, 6.G.A.3

- ✓ Show What You Know **531**
-   Vocabulary Builder **532**
- 1 Algebra • Area of Parallelograms **533**
- 2 Investigate • Explore Area of Triangles **539**
- 3 Algebra • Area of Triangles **545**
- 4 Investigate • Explore Area of Trapezoids **551**
- 5 Algebra • Area of Trapezoids **557**
- ✓ Mid-Chapter Checkpoint **563**
- 6 Area of Regular Polygons **565**
- 7 Composite Figures **571**
- 8 Problem Solving • Changing Dimensions **577**
- 9 Figures on the Coordinate Plane **583**
- ✓ Chapter 10 Review/Test **589**

11 Surface Area and Volume 595
Domain Geometry
COMMON CORE STATE STANDARDS 6.G.A.2, 6.G.A.4

- ✓ Show What You Know **595**
-   Vocabulary Builder **596**
- 1 Three-Dimensional Figures and Nets **597**
- 2 Investigate • Explore Surface Area Using Nets **603**
- 3 Algebra • Surface Area of Prisms **609**
- 4 Algebra • Surface Area of Pyramids **615**
- ✓ Mid-Chapter Checkpoint **621**
- 5 Investigate • Fractions and Volume **623**
- 6 Algebra • Volume of Rectangular Prisms **629**
- 7 Problem Solving • Geometric Measurements **635**
- ✓ Chapter 11 Review/Test **641**

Critical Area

GO DIGITAL

Go online! Your math lessons are interactive. Use iTools, Animated Math Models, the Multimedia eGlossary, and more.

Chapter 10 Overview
In this chapter, you will explore and discover answers to the following **Essential Questions**:
- How can you use measurements to describe two-dimensional figures?
- What does area represent?
- How are the areas of rectangles and parallelograms related?
- How are the areas of triangles and trapezoids related?

Chapter 11 Overview
In this chapter, you will explore and discover answers to the following **Essential Questions**:
- How can you use measurements to describe three-dimensional figures?
- How can you use a net to find the surface area of a three-dimensional figure?
- How can you find the volume of a rectangular prism?

Chapter 12 Overview

In this chapter, you will explore and discover answers to the following **Essential Questions**:

- How can you display data and analyze measures of center?
- When does it make sense to display data in a dot plot? in a histogram?
- What are the differences between the three measures of center?

Practice and Homework

Lesson Check and Spiral Review in every lesson

Chapter 13 Overview

In this chapter, you will explore and discover answers to the following **Essential Questions**:

- How can you describe the shape of a data set using graphs, measures of center, and measures of variability?
- How do you calculate the different measures of center?
- How do you calculate the different measures of variability?

Data Displays and Measures of Center — 647

Domain Statistics and Probability
COMMON CORE STATE STANDARDS 6.SP.A.1, 6.SP.B.4, 6.SP.B.5a, 6.SP.B.5b, 6.SP.B.5c, 6.SP.B.5d

✓ Show What You Know 647
   Vocabulary Builder 648
1 Recognize Statistical Questions 649
2 Describe Data Collection 655
3 Dot Plots and Frequency Tables 661
4 Histograms 667
✓ Mid-Chapter Checkpoint 673
5 **Investigate** • Mean as Fair Share and Balance Point 675
6 Measures of Center 681
7 Effects of Outliers 687
8 **Problem Solving** • Data Displays 693
✓ Chapter 12 Review/Test 699

Variability and Data Distributions — 705

Domain Statistics and Probability
COMMON CORE STATE STANDARDS 6.SP.A.2, 6.SP.A.3, 6.SP.B.4, 6.SP.B.5c, 6.SP.B.5d

✓ Show What You Know 705
   Vocabulary Builder 706
1 Patterns in Data 707
2 Box Plots 713
3 **Investigate** • Mean Absolute Deviation 719
4 Measures of Variability 725
✓ Mid-Chapter Checkpoint 731
5 Choose Appropriate Measures of Center and Variability 733
6 Apply Measures of Center and Variability 739
7 Describe Distributions 745
8 **Problem Solving** • Misleading Statistics 751
✓ Chapter 13 Review/Test 757

   Glossary H1
   Common Core State Standards Correlations H17
   Index H27
   Table of Measures H39

Glossary

Pronunciation Key

a	add, map	f	fit, half	n	nice, tin	p	pit, stop
ā	ace, rate	g	go, log	ng	ring, song	r	run, poor
â(r)	care, air	h	hope, hate	o	odd, hot	s	see, pass
ä	palm, father	i	it, give	ō	open, so	sh	sure, rush
b	bat, rub	ī	ice, write	ô	order, jaw	t	talk, sit
ch	check, catch	j	joy, ledge	oi	oil, boy	th	thin, both
d	dog, rod	k	cool, take	ou	pout, now	<u>th</u>	this, bathe
e	end, pet	l	look, rule	o͞o	took, full	u	up, done
ē	equal, tree	m	move, seem	o͞o	pool, food	ù	pull book

û(r)	burn, term	
yo͞o	fuse, few	
v	vain, eve	
w	win, away	
y	yet, yearn	
z	zest, muse	
zh	vision, pleasure	

ə the schwa, an unstressed vowel representing the sound spelled *a* in *a*bove, *e* in sick*e*n, *i* in poss*i*ble, *o* in mel*o*n, *u* in circ*u*s

Other symbols:
• separates words into syllables
′ indicates stress on a syllable

absolute value [ab′sə•lo͞ot val′yo͞o] **valor absoluto** The distance of an integer from zero on a number line (p. 165)

acute angle [ə•kyo͞ot′ ang′gəl] **ángulo agudo** An angle that has a measure less than a right angle (less than 90° and greater than 0°)
Example:

acute triangle [ə•kyo͞ot′ trī′ang•gəl] **triángulo acutángulo** A triangle that has three acute angles

addend [ad′end] **sumando** A number that is added to another in an addition problem

addition [ə•dish′ən] **suma** The process of finding the total number of items when two or more groups of items are joined; the inverse operation of subtraction

Addition Property of Equality [ə•dish′ən präp′ər•tē əv ē•kwôl′ə•tē] **propiedad de suma de la igualdad** The property that states that if you add the same number to both sides of an equation, the sides remain equal

additive inverse [ad′ə•tiv in′v ûrs] **inverso aditivo** The number which, when added to the given number, equals zero

algebraic expression [al•jə•brā′ik ek•spresh′ən] **expresión algebraica** An expression that includes at least one variable (p. 369)
Examples: $x + 5$, $3a - 4$

angle [ang′gəl] **ángulo** A shape formed by two rays that share the same endpoint
Example:

area [âr′ē•ə] **área** The measure of the number of unit squares needed to cover a surface (p. 533)

Glossary **H1**

array [ə•rā′] **matriz** An arrangement of objects in rows and columns
Example:

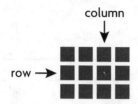

Associative Property of Addition [ə•sō′shē•ə•āt•iv präp′ər•tē əv ə•dish′ən] **propiedad asociativa de la suma** The property that states that when the grouping of addends is changed, the sum is the same
Example: $(5 + 8) + 4 = 5 + (8 + 4)$

Associative Property of Multiplication [ə•sō′shē•ə•tiv präp′ər•tē əv mul•tə•pli•kā′shən] **propiedad asociativa de la multiplicación** The property that states that when the grouping of factors is changed, the product is the same
Example: $(2 \times 3) \times 4 = 2 \times (3 \times 4)$

bar graph [bär graf] **gráfica de barras** A graph that uses horizontal or vertical bars to display countable data
Example:

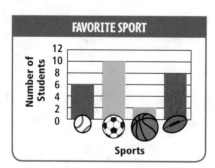

base [bās] (arithmetic) **base** A number used as a repeated factor (p. 357)
Example: $8^3 = 8 \times 8 \times 8$. The base is 8.

base [bās] (geometry) **base** In two dimensions, one side of a triangle or parallelogram which is used to help find the area. In three dimensions, a plane figure, usually a polygon or circle, which is used to partially describe a solid figure and to help find the volume of some solid figures. See also *height*.
Examples:

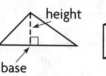

benchmark [bench′märk] **punto de referencia** A familiar number used as a point of reference

billion [bil′yən] **millardo** 1,000 millions; written as 1,000,000,000

box plot [bäks plät] **diagrama de caja** A graph that shows how data are distributed using the median, quartiles, least value, and greatest value (p. 714)
Example:

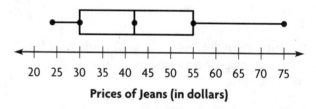

capacity [kə•pas′i•tē] **capacidad** The amount a container can hold (p. 321)
Examples: $\frac{1}{2}$ gallon, 2 quarts

Celsius (°C) [sel′sē•əs] **Celsius (°C)** A metric scale for measuring temperature

closed figure [klōzd fig′yər] **figura cerrada** A figure that begins and ends at the same point

coefficient [kō•ə•fish′ənt] **coeficiente** A number that is multiplied by a variable (p. 376)
Example: 6 is the coefficient of *x* in 6*x*

common denominator [käm′ən dē•näm′ə•nāt•ər] **denominador común** A common multiple of two or more denominators
Example: Some common denominators for $\frac{1}{4}$ and $\frac{5}{6}$ are 12, 24, and 36.

common factor [käm′ən fak′tər] **factor común** A number that is a factor of two or more numbers (p. 23)

common multiple [käm′ən mul′tə•pəl] **múltiplo común** A number that is a multiple of two or more numbers

Commutative Property of Addition [kə•myōōt′ ə•tiv präp′ər•tē əv ə•dish′ən] **propiedad conmutativa de la suma** The property that states that when the order of two addends is changed, the sum is the same
Example: $4 + 5 = 5 + 4$

Commutative Property of Multiplication [kə•myōōt′ə•tiv präp′ər•tē əv mul•tə•pli•kāsh′ən] **propiedad conmutativa de la multiplicación** The property that states that when the order of two factors is changed, the product is the same
Example: $4 \times 5 = 5 \times 4$

compatible numbers [kəm•pat′ə•bəl num′bərz] **números compatibles** Numbers that are easy to compute with mentally

composite figure [kəm•päz′it fig′yər] **figura compuesta** A figure that is made up of two or more simpler figures, such as triangles and quadrilaterals (p. 571)
Example:

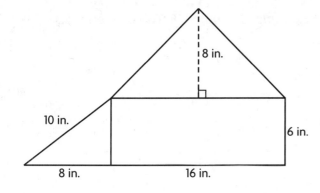

composite number [kəm•päz′it num′bər] **número compuesto** A number having more than two factors
Example: 6 is a composite number, since its factors are 1, 2, 3, and 6.

cone [kōn] **cono** A solid figure that has a flat, circular base and one vertex
Example:

congruent [kən•grōō′ənt] **congruente** Having the same size and shape (p. 539)
Example:

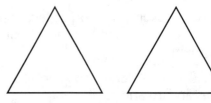

conversion factor [kən•vûr′zhən fak′tər] **factor de conversión** A rate in which two quantities are equal, but use different units (p. 315)

coordinate plane [kō•ôrd′n•it plān] **plano cartesiano** A plane formed by a horizontal line called the *x*-axis and a vertical line called the *y*-axis (p. 177)
Example:

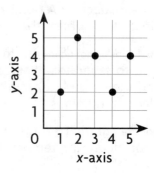

Glossary H3

cube [kyoob] **cubo** A solid figure with six congruent square faces
Example:

cubic unit [kyoo′bik yoo′nit] **unidad cúbica** A unit used to measure volume such as cubic foot (ft³), cubic meter (m³), and so on

data [dāt′ə] **datos** Information collected about people or things, often to draw conclusions about them (p. 649)

decagon [dek′ə·gän] **decágono** A polygon with 10 sides and 10 angles
Examples:

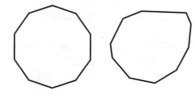

decimal [des′ə·məl] **decimal** A number with one or more digits to the right of the decimal point

decimal point [des′ə·məl point] **punto decimal** A symbol used to separate dollars from cents in money, and the ones place from the tenths place in decimal numbers

degree (°) [di·grē′] **grado (°)** A unit for measuring angles or for measuring temperature

degree Celsius (°C) [di·grē′ sel′sē·əs] **grado Celsius** A metric unit for measuring temperature

degree Fahrenheit (°F) [di·grē′ făr′ən·hīt] **grado Fahrenheit** A customary unit for measuring temperature

denominator [de·näm′ə·nāt·ər] **denominador** The number below the bar in a fraction that tells how many equal parts are in the whole or in the group
Example: $\frac{3}{4}$ ← denominator

dependent variable [de·pen′dənt vâr′ē·ə·bəl] **variable dependiente** A variable whose value depends on the value of another quantity (p. 491)

difference [dif′ər·əns] **diferencia** The answer to a subtraction problem

digit [dij′it] **dígito** Any one of the ten symbols 0, 1, 2, 3, 4, 5, 6, 7, 8, 9 used to write numbers

dimension [də·men′shən] **dimensión** A measure in one direction

distribution [dis·tri·byoo′shən] **distribución** The overall shape of a data set

Distributive Property [di·strib′yoo·tiv präp′ər·tē] **propiedad distributiva** The property that states that multiplying a sum by a number is the same as multiplying each addend in the sum by the number and then adding the products (p. 24)
Example: $3 \times (4 + 2) = (3 \times 4) + (3 \times 2)$
$3 \times 6 = 12 + 6$
$18 = 18$

divide [də·vīd′] **dividir** To separate into equal groups; the inverse operation of multiplication

dividend [div′ə·dend] **dividendo** The number that is to be divided in a division problem
Example: $36 \div 6$; $6\overline{)36}$ The dividend is 36.

H4 Glossary

divisible [də•viz′ə•bəl] **divisible** A number is divisible by another number if the quotient is a counting number and the remainder is zero
Example: 18 is divisible by 3.

division [də•vizh′ən] **división** The process of sharing a number of items to find how many groups can be made or how many items will be in a group; the operation that is the inverse of multiplication

Division Property of Equality [də•vizh′ən präp′ər•tē əv ē•kwôl′ə•tē] **propiedad de división de la igualdad** The property that states that if you divide both sides of an equation by the same nonzero number, the sides remain equal

divisor [də•vī′zər] **divisor** The number that divides the dividend
Example: $15 \div 3$; $3\overline{)15}$ The divisor is 3.

dot plot [dot plät] **diagrama de puntos** A graph that shows frequency of data along a number line (p. 661)
Example:

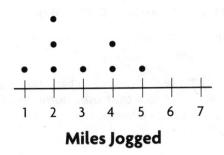

Miles Jogged

edge [ej] **arista** The line where two faces of a solid figure meet
Example:

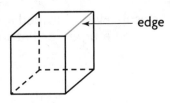

equation [i•kwā′zhən] **ecuación** An algebraic or numerical sentence that shows that two quantities are equal (p. 421)

equilateral triangle [ē•kwi•lat′ər•əl trī′ang•gəl] **triángulo equilátero** A triangle with three congruent sides
Example:

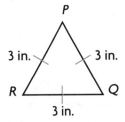

equivalent [ē•kwiv′ə•lənt] **equivalente** Having the same value

equivalent decimals [ē•kwiv′ə•lənt des′ə•məlz] **decimales equivalentes** Decimals that name the same number or amount
Example: 0.4 = 0.40 = 0.400

equivalent expressions [ē•kwiv′ə•lənt ek•spresh′ənz] **expresiones equivalentes** Expressions that are equal to each other for any values of their variables (p. 401)
Example: $2x + 4x = 6x$

equivalent fractions [ē•kwiv′ə•lənt frak′shənz] **fracciones equivalentes** Fractions that name the same amount or part
Example: $\frac{3}{4} = \frac{6}{8}$

equivalent ratios [ē•kwiv′ə•lənt rā′shē•ōz] **razones equivalentes** Ratios that name the same comparison (p. 223)

estimate [es′tə•mit] *noun* **estimación (s)** A number close to an exact amount

estimate [es′tə•māt] *verb* **estimar (v)** To find a number that is close to an exact amount

evaluate [ē·val′yōō·āt] **evaluar** To find the value of a numerical or algebraic expression (p. 363)

even [ē′vən] **par** A whole number that has a 0, 2, 4, 6, or 8 in the ones place

expanded form [ek·span′did fôrm] **forma desarrollada** A way to write numbers by showing the value of each digit
Example: 832 = 800 + 30 + 2

exponent [eks′pōn·ənt] **exponente** A number that shows how many times the base is used as a factor (p. 357)
Example: $10^3 = 10 \times 10 \times 10$; 3 is the exponent.

> **Word History**
>
> *Exponent* comes from the combination of the Latin roots *ex* ("out of") + *ponere* ("to place"). In the 17th century, mathematicians began to use complicated quantities. The idea of positioning a number by raising it "out of place" is traced to René Descartes.

expression [ek·spresh′ən] **expresión** A mathematical phrase or the part of a number sentence that combines numbers, operation signs, and sometimes variables, but does not have an equal or inequality sign

face [fās] **cara** A polygon that is a flat surface of a solid figure
Example:

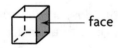

fact family [fakt fam′ə·lē] **familia de operaciones** A set of related multiplication and division, or addition and subtraction, equations
Example: $7 \times 8 = 56$; $8 \times 7 = 56$; $56 \div 7 = 8$; $56 \div 8 = 7$

factor [fak′tər] **factor** A number multiplied by another number to find a product

factor tree [fak′tər trē] **árbol de factores** A diagram that shows the prime factors of a number
Example:

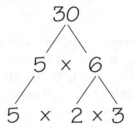

Fahrenheit (°F) [fâr′ən·hīt] **Fahrenheit (°F)** A customary scale for measuring temperature

formula [fôr′myōō·lə] **fórmula** A set of symbols that expresses a mathematical rule
Example: $A = b \times h$

fraction [frak′shən] **fracción** A number that names a part of a whole or a part of a group

frequency [frē′kwən·sē] **frecuencia** The number of times an event occurs (p. 661)

frequency table [frē′kwən·sē tā′bəl] **tabla de frecuencia** A table that uses numbers to record data about how often an event occurs (p. 662)

greatest common factor (GCF) [grāt′est käm′ən fak′tər] **máximo común divisor (MCD)** The greatest factor that two or more numbers have in common (p. 23)
Example: 6 is the GCF of 18 and 30.

grid [grid] **cuadrícula** Evenly divided and equally spaced squares on a figure or flat surface

H

height [hīt] **altura** The length of a perpendicular from the base to the top of a plane figure or solid figure
Example:

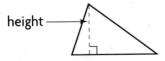

hexagon [hek′sə•gän] **hexágono** A polygon with six sides and six angles
Examples:

histogram [his′tə•gram] **histograma** A type of bar graph that shows the frequencies of data in intervals. (p. 667)
Example:

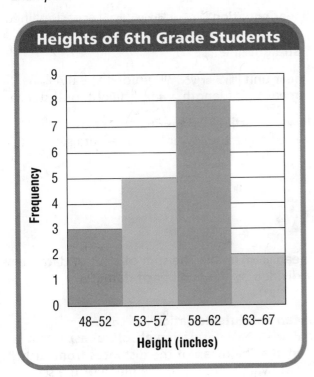

horizontal [hôr•i•zänt′əl] **horizontal** Extending left and right

hundredth [hun′drədth] **centésimo** One of one hundred equal parts
Examples: 0.56, $\frac{56}{100}$, fifty-six hundredths

I

Identity Property of Addition [ī•den′tə•tē präp′ər•tē əv ə•dish′ən] **propiedad de identidad de la suma** The property that states that when you add zero to a number, the result is that number

Identity Property of Multiplication [ī•den′tə•tē präp′ər•tē əv mul•tə•pli•kāsh′ən] **propiedad de identidad de la multiplicación** The property that states that the product of any number and 1 is that number

independent variable [in•dē•pen′dənt′ vâr′ē•ə•bəl] **variable independiente** A variable whose value determines the value of another quantity (p. 491)

inequality [in•ē•kwôl′ə•tē] **desigualdad** A mathematical sentence that contains the symbol <, >, ≤, ≥, or ≠ (p. 465)

integers [in′tə•jərz] **enteros** The set of whole numbers and their opposites (p. 139)

interquartile range [in′tûr•kwôr′tīl rānj] **rango intercuartil** The difference between the upper and lower quartiles of a data set (p. 726)

intersecting lines [in•tər•sekt′ing līnz] **líneas secantes** Lines that cross each other at exactly one point
Example:

inverse operations [in′vûrs äp•pə•rā′shənz] **operaciones inversas** Opposite operations, or operations that undo each other, such as addition and subtraction or multiplication and division (p. 439)

Glossary H7

key [kē] **clave** The part of a map or graph that explains the symbols

kite [kīt] **cometa** A quadrilateral with exactly two pairs of congruent sides that are next to each other; no two sides are parallel
Example:

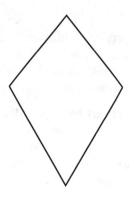

ladder diagram [lad′ər dī′ə•gram] **diagrama de escalera** A diagram that shows the steps of repeatedly dividing by a prime number until the quotient is 1

lateral area [lat′ər•əl âr′ē•ə] **área lateral** The sum of the areas of the lateral faces of a solid (p. 616)

lateral face [lat′ər•əl fās] **cara lateral** Any surface of a polyhedron other than a base

least common denominator (LCD) [lēst käm′ən dē•näm′ə•nāt•ər] **mínimo común denominador (m.c.d.)** The least common multiple of two or more denominators
Example: The LCD for $\frac{1}{4}$ and $\frac{5}{6}$ is 12.

least common multiple (LCM) [lēst käm′ən mul′tə•pəl] **mínimo común múltiplo (m.c.m.)** The least number that is a common multiple of two or more numbers (p. 17)

like terms [līk tûrmz] **términos semejantes** Expressions that have the same variable with the same exponent (p. 395)

line [līn] **línea** A straight path in a plane, extending in both directions with no endpoints
Example:

◄――――――――►

line graph [līn graf] **gráfica lineal** A graph that uses line segments to show how data change over time

line of symmetry [līn əv sim′ə•trē] **eje de simetría** A line that divides a figure into two halves that are reflections of each other (p. 184)

line segment [līn seg′mənt] **segmento** A part of a line that includes two points called endpoints and all the points between them
Example:

●―――――――●

line symmetry [līn sim′ə•trē] **simetría axial** A figure has line symmetry if it can be folded about a line so that its two parts match exactly. (p. 184)

linear equation [lin′ē•ər ē•kwā′zhən] **ecuación lineal** An equation that, when graphed, forms a straight line (p. 517)

linear unit [lin′ē•ər yōō′nit] **unidad lineal** A measure of length, width, height, or distance

lower quartile [lō′ər kwôr′tīl] **primer cuartil** The median of the lower half of a data set (p. 713)

mean [mēn] **media** The sum of a set of data items divided by the number of data items (p. 681)

mean absolute deviation [mēn ab′sə•lōōt dē•vē•ā′shən] **desviación absoluta respecto a la media** The mean of the distances from each data value in a set to the mean of the set (p. 720)

measure of center [mezh′ər əv sent′ər] **medida de tendencia central** A single value used to describe the middle of a data set (p. 681)
Examples: mean, median, mode

measure of variability [mezh′ər əv vâr′ē•ə•bil′ə•tē] **medida de dispersión** A single value used to describe how the values in a data set are spread out (p. 725)
Examples: range, interquartile range, mean absolute deviation

median [mē′dēən] **mediana** The middle value when a data set is written in order from least to greatest, or the mean of the two middle values when there is an even number of items (p. 681)

midpoint [mid′point] **punto medio** A point on a line segment that is equally distant from either endpoint

million [mil′yən] **millón** 1,000 thousands; written as 1,000,000

mixed number [mikst num′bər] **número mixto** A number that is made up of a whole number and a fraction
Example: $1\frac{5}{8}$

mode [mōd] **moda** The value(s) in a data set that occurs the most often (p. 681)

multiple [mul′tə•pəl] **múltiplo** The product of two counting numbers is a multiple of each of those numbers

multiplication [mul•tə•pli•kā′shən] **multiplicación** A process to find the total number of items made up of equal-sized groups, or to find the total number of items in a given number of groups; It is the inverse operation of division.

Multiplication Property of Equality [mul•tə•pli•kā′shən präp′ər•tē əv ē•kwôl′ə•tē] **propiedad de multiplicación de la igualdad** The property that states that if you multiply both sides of an equation by the same number, the sides remain equal

multiplicative inverse [mul′tə•pli•kāt•iv in′vûrs] **inverso multiplicativo** A reciprocal of a number that is multiplied by that number resulting in a product of 1 (p. 108)

multiply [mul′tə•plī] **multiplicar** When you combine equal groups, you can multiply to find how many in all; the inverse operation of division

negative integer [neg′ə•tiv in′tə•jər] **entero negativo** Any integer less than zero
Examples: $^-4$, $^-5$, and $^-6$ are negative integers.

net [net] **plantilla** A two-dimensional pattern that can be folded into a three-dimensional polyhedron (p. 597)
Example:

not equal to (≠) [not ē′kwəl too] **no igual a** A symbol that indicates one quantity is not equal to another

number line [num′bər līn] **recta numérica** A line on which numbers can be located
Example:

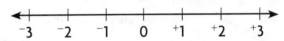

numerator [noo′mər•āt•ər] **numerador** The number above the bar in a fraction that tells how many equal parts of the whole are being considered
Example: $\frac{3}{4}$ ← numerator

numerical expression [noo•mer′i•kəl ek•spresh′ən] **expresión numérica** A mathematical phrase that uses only numbers and operation signs (p. 363)

O

obtuse angle [äb•tōōs' ang'gəl] **ángulo obtuso** An angle whose measure is greater than 90° and less than 180°
Example:

obtuse triangle [äb•tōōs' trī'ang•gəl] **triángulo obtusángulo** A triangle that has one obtuse angle

octagon [äk'tə•gän] **octágono** A polygon with eight sides and eight angles
Examples:

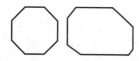

odd [od] **impar** A whole number that has a 1, 3, 5, 7, or 9 in the ones place

open figure [ō'pən fig'yər] **figura abierta** A figure that does not begin and end at the same point

opposites [äp'ə•zits] **opuestos** Two numbers that are the same distance, but in opposite directions, from zero on a number line (p. 139)

order of operations [ôr'dər əv äp•ə•rā'shənz] **orden de las operaciones** A special set of rules which gives the order in which calculations are done in an expression (p. 363)

ordered pair [ôr'dərd pâr] **par ordenado** A pair of numbers used to locate a point on a grid. The first number tells the left-right position and the second number tells the up-down position. (p. 177)

origin [ôr'ə•jin] **origen** The point where the two axes of a coordinate plane intersect; (0,0) (p. 177)

outlier [out'lī•ər] **valor atípico** A value much higher or much lower than the other values in a data set (p. 687)

overestimate [ō'vər•es•tə•mit] **sobrestimar** An estimate that is greater than the exact answer

P

parallel lines [pâr'ə•lel līnz] **líneas paralelas** Lines in the same plane that never intersect and are always the same distance apart
Example:

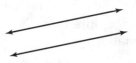

parallelogram [pâr•ə•lel'ə•gram] **paralelogramo** A quadrilateral whose opposite sides are parallel and congruent
Example:

parentheses [pə•ren'thə•sēz] **paréntesis** The symbols used to show which operation or operations in an expression should be done first

partial product [pär'shəl präd'əkt] **producto parcial** A method of multiplying in which the ones, tens, hundreds, and so on are multiplied separately and then the products are added together

pattern [pat'ərn] **patrón** An ordered set of numbers or objects; the order helps you predict what will come next
Examples: 2, 4, 6, 8, 10

pentagon [pen'tə•gän] **pentágono** A polygon with five sides and five angles
Examples:

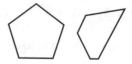

H10 Glossary

percent [pər•sent′] **porcentaje** The comparison of a number to 100; percent means "per hundred" (p. 269)

perimeter [pə•rim′ə•tər] **perímetro** The distance around a closed plane figure

period [pir′ē•əd] **período** Each group of three digits separated by commas in a multidigit number
Example: 85,643,900 has three periods.

perpendicular lines [pər•pən•dik′yōō•lər līnz] **líneas perpendiculares** Two lines that intersect to form four right angles
Example:

pictograph [pik′tə•graf] **pictografía** A graph that displays countable data with symbols or pictures
Example:

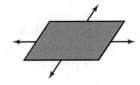

place value [plās val′yōō] **valor posicional** The value of each digit in a number based on the location of the digit

plane [plān] **plano** A flat surface that extends without end in all directions
Example:

plane figure [plān fig′yər] **figura plana** A figure that lies in a plane; a figure having length and width

point [point] **punto** An exact location in space

polygon [päl′i•gän] **polígono** A closed plane figure formed by three or more line segments
Examples:

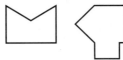

Polygons   Not Polygons

polyhedron [päl•i•hē′drən] **poliedro** A solid figure with faces that are polygons
Examples:

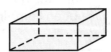

positive integer [päz′ə•tiv in′tə•jər] **entero positivo** Any integer greater than zero

prime factor [prīm fak′tər] **factor primo** A factor that is a prime number

prime factorization [prīm fak•tər•ə•zā′shən] **descomposición en factores primos** A number written as the product of all its prime factors (p. 11)

prime number [prīm num′bər] **número primo** A number that has exactly two factors: 1 and itself
Examples: 2, 3, 5, 7, 11, 13, 17, and 19 are prime numbers. 1 is not a prime number.

prism [priz′əm] **prisma** A solid figure that has two congruent, polygon-shaped bases, and other faces that are all rectangles
Examples:

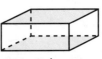

rectangular prism   triangular prism

product [präd′əkt] **producto** The answer to a multiplication problem

pyramid [pir′ə•mid] **pirámide** A solid figure with a polygon base and all other faces as triangles that meet at a common vertex
Example:

> **Word History**
>
> A fire is sometimes in the shape of a pyramid, with a point at the top and a wider base. This may be how *pyramid* got its name. The Greek word for fire was *pura*, which may have been combined with the Egyptian word *mer*.

Q

quadrants [kwä′drənts] **cuadrantes** The four regions of the coordinate plane separated by the *x*- and *y*-axes (p. 183)

quadrilateral [kwä•dri•lat′ər•əl] **cuadrilátero** A polygon with four sides and four angles
Example:

quotient [kwō′shənt] **cociente** The number that results from dividing
Example: 8 ÷ 4 = 2. The quotient is 2.

R

range [rānj] **rango** The difference between the greatest and least numbers in a data set (p. 726)

rate [rāt] **tasa** A ratio that compares two quantities having different units of measure (p. 218)

ratio [rā′shē•ō] **razón** A comparison of two numbers, *a* and *b*, that can be written as a fraction $\frac{a}{b}$ (p. 211)

rational number [rash′•ən•əl num′bər] **número racional** Any number that can be written as a ratio $\frac{a}{b}$ where *a* and *b* are integers and $b \neq 0$. (p. 151)

ray [rā] **semirrecta** A part of a line; it has one endpoint and continues without end in one direction
Example:

reciprocal [ri•sip′rə•kəl] **recíproco** Two numbers are reciprocals of each other if their product equals 1. (p. 108)

rectangle [rek′tang•gəl] **rectángulo** A parallelogram with four right angles
Example:

rectangular prism [rek•tang′gyə•lər priz′əm] **prisma rectangular** A solid figure in which all six faces are rectangles
Example:

reflection [ri•flek′shən] **reflexión** A movement of a figure to a new position by flipping it over a line; a flip
Example:

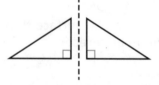

regroup [rē•grōōp′] **reagrupar** To exchange amounts of equal value to rename a number
Example: 5 + 8 = 13 ones or 1 ten 3 ones

regular polygon [reg′yə•lər päl′i•gän] **polígono regular** A polygon in which all sides are congruent and all angles are congruent (p. 565)

relative frequency table [rel′ə•tiv frē′kwən•sē tā′bəl] **tabla de frecuencia relativa** A table that shows the percent of time each piece of data occurs (p. 662)

remainder [ri•mān′dər] **residuo** The amount left over when a number cannot be divided equally

rhombus [räm′bəs] **rombo** A parallelogram with four congruent sides
Example:

Word History

Rhombus is almost identical to its Greek origin, *rhombos*. The original meaning was "spinning top" or "magic wheel," which is easy to imagine when you look at a rhombus, an equilateral parallelogram.

right triangle [rīt trī′ang•gəl] **triángulo rectángulo** A triangle that has a right angle
Example:

round [round] **redondear** To replace a number with one that is simpler and is approximately the same size as the original number
Example: 114.6 rounded to the nearest ten is 110 and to the nearest unit is 115.

sequence [sē′kwəns] **secuencia** An ordered set of numbers

simplest form [sim′pləst fôrm] **mínima expresión** A fraction is in simplest form when the numerator and denominator have only 1 as a common factor

simplify [sim′plə•fī] **simplificar** The process of dividing the numerator and denominator of a fraction or ratio by a common factor

solid figure [sä′lid fig′yər] **cuerpo geométrico** A three-dimensional figure having length, width, and height (p. 597)

solution of an equation [sə•loo′shən əv an ē•kwā′zhən] **solución de una ecuación** A value that, when substituted for the variable, makes an equation true (p. 421)

solution of an inequality [sə•loo′shən əv an in•ē•kwôl′ə•tē] **solución de una desigualdad** A value that, when substituted for the variable, makes an inequality true (p. 465)

square [skwâr] **cuadrado** A polygon with four equal, or congruent, sides and four right angles

square pyramid [skwâr pir′ə•mid] **pirámide cuadrada** A solid figure with a square base and with four triangular faces that have a common vertex
Example:

square unit [skwâr yoo′nit] **unidad cuadrada** A unit used to measure area such as square foot (ft^2), square meter (m^2), and so on

standard form [stan′dərd fôrm] **forma normal** A way to write numbers by using the digits 0–9, with each digit having a place value
Example: 456 ← standard form

statistical question [stə·tis′ti·kəl kwes′chən] **pregunta estadística** A question that asks about a set of data that can vary (p. 649)
Example: How many desks are in each classroom in my school?

Substitution Property of Equality [sub·stə·tōō′shən präp′ər·tē əv ē·kwôl′ə·tē] **propiedad de sustitución de la igualdad** The property that states that if you have one quantity equal to another, you can substitute that quantity for the other in an equation

subtraction [səb·trak′shən] **resta** The process of finding how many are left when a number of items are taken away from a group of items; the process of finding the difference when two groups are compared; the inverse operation of addition

Subtraction Property of Equality [səb·trak′shən präp′ər·tē əv ē·kwôl′ə·tē] **propiedad de resta de la igualdad** The property that states that if you subtract the same number from both sides of an equation, the sides remain equal

sum [sum] **suma o total** The answer to an addition problem

surface area [sûr′fis âr′ē·ə] **área total** The sum of the areas of all the faces, or surfaces, of a solid figure (p. 603)

tally table [tal′ē tā′bəl] **tabla de conteo** A table that uses tally marks to record data

tenth [tenth] **décimo** One of ten equal parts
Example: 0.7 = seven tenths

terms [tûrmz] **términos** The parts of an expression that are separated by an addition or subtraction sign (p. 376)

thousandth [thou′zəndth] **milésimo** One of one thousand equal parts
Example: 0.006 = six thousandths

three-dimensional [thrē də·men′shə·nəl] **tridimensional** Measured in three directions, such as length, width, and height

three-dimensional solid [thrē də·men′shə·nəl säl′id] **figura tridimensional** See *solid figure*

trapezoid [trap′i·zoid] **trapecio** A quadrilateral with at least one pair of parallel sides (p. 551)
Examples:

tree diagram [trē dī′ə·gram] **diagrama de árbol** A branching diagram that shows all possible outcomes of an event

trend [trend] **tendencia** A pattern over time, in all or part of a graph, where the data increase, decrease, or stay the same

triangle [trī′ang·gəl] **triángulo** A polygon with three sides and three angles
Examples:

triangular prism [trī·ang′gyə·lər priz′əm] **prisma triangular** A solid figure that has two triangular bases and three rectangular faces

two-dimensional [tōō də·men′shə·nəl] **bidimensional** Measured in two directions, such as length and width

two-dimensional figure [tōō də·men′shə·nəl fig′yər] **figura bidimensional** See *plane figure*

underestimate [un•dər•es′tə•mit] **subestimar** An estimate that is less than the exact answer

unit cube [yōō′nit kyōōb] **cubo unitaria** A cube that has a length, width, and height of 1 unit

unit fraction [yōō′nit frak′shən] **fraccion unitaria** A fraction that has 1 as a numerator

unit rate [yōō′nit rāt] **tasa por unidad** A rate expressed so that the second term in the ratio is one unit (p. 218)
Example: 55 mi per hr

unit square [yōō′nit skwâr] **cuadrado de una unidad** A square with a side length of 1 unit, used to measure area

unlike fractions [un′līk frak′shənz] **fracciones no semejantes** Fractions with different denominators

upper quartile [up′ər kwôr′tīl] **tercer cuartil** The median of the upper half of a data set (p. 713)

variable [vâr′ē•ə•bəl] **variable** A letter or symbol that stands for an unknown number or numbers (p. 369)

Venn diagram [ven dī′ə•gram] **diagrama de Venn** A diagram that shows relationships among sets of things
Example:

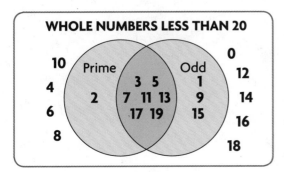

vertex [vûr′teks] **vértice** The point where two or more rays meet; the point of intersection of two sides of a polygon; the point of intersection of three (or more) edges of a solid figure; the top point of a cone; the plural of *vertex* is *vertices*
Examples:

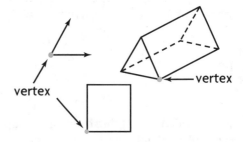

vertical [vûr′ti•kəl] **vertical** Extending up and down

volume [väl′yōōm] **volumen** The measure of the space a solid figure occupies (p. 623)

weight [wāt] **peso** How heavy an object is

whole number [hōl num′bər] **número entero** One of the numbers 0, 1, 2, 3, 4, . . . ; the set of whole numbers goes on without end

x-axis [eks ak′sis] **eje de la *x*** The horizontal number line on a coordinate plane (p. 177)

x-coordinate [eks kō•ôrd′n•it] **coordenada *x*** The first number in an ordered pair; tells the distance to move right or left from (0,0) (p. 177)

y-axis [wī ak′sis] **eje de la y** The vertical number line on a coordinate plane (p. 177)

y-coordinate [wī kō•ôrd′•n•it] **coordenada y** The second number in an ordered pair; tells the distance to move up or down from (0,0) (p. 177)

Zero Property of Multiplication [zē′rō präp′ər•tē əv mul•tə•pli•kā′shən] **propiedad del cero de la multiplicación** The property that states that when you multiply by zero, the product is zero

Correlations

 COMMON CORE STATE STANDARDS

Standards You Will Learn

Mathematical Practices		Some examples are:
MP1	Make sense of problems and persevere in solving them.	Lessons 1.1, 2.6, 2.9, 6.3, 6.5, 7.6, 8.7, 12.8, 13.2, 13.7
MP2	Reason abstractly and quantitatively.	Lessons 1.1, 2.3, 3.1, 7.3, 7.4, 7.9, 11.2, 12.7, 13.5
MP3	Construct viable arguments and critique the reasoning of others.	Lessonss 1.2, 2.3, 3.5, 6.4, 7.7, 8.1, 11.4, 12.5, 13.7
MP4	Model with mathematics.	Lessons 1.4, 2.5, 2.8, 6.3, 7.2, 8.2, 10.4, 12.4, 13.3
MP5	Use appropriate tools strategically.	Lessons 2.8, 3.4, 5.1, 6.3, 8.3, 9.2, 12.3, 12.8, 13.7
MP6	Attend to precision.	Lessons 1.6, 2.9, 3.5, 7.4, 7.9, 8.1, 13.6, 13.7, 13.8
MP7	Look for and make use of structure.	Lessons 1.2, 2.9, 3.1, 5.2, 6.5, 8.6, 13.1, 13.4, 13.8
MP8	Look for and express regularity in repeated reasoning.	Lessons 1.9, 2.7, 3.2, 4.5, 6.1, 7.1, 10.2, 12.5, 13.1
Domain: Ratios and Proportional Relationships		**Student Edition Lessons**
Understand ratio concepts and use ratio reasoning to solve problems.		
6.RP.A.1	Understand the concept of a ratio and use ratio language to describe a ratio relationship between two quantities.	Lessons 4.1, 4.2
6.RP.A.2	Understand the concept of a unit rate a/b associated with a ratio $a:b$ with $b \neq 0$, and use rate language in the context of a ratio relationship.	Lessons 4.2, 4.6

Correlations H17

Standards You Will Learn

Student Edition Lessons

Domain: Ratios and Proportional Relationships (Continued)

6.RP.A.3	Use ratio and rate reasoning to solve real-world and mathematical problems, e.g., by reasoning about tables of equivalent ratios, tape diagrams, double number line diagrams, or equations.	
	a. Make tables of equivalent ratios relating quantities with whole-number measurements, find missing values in the tables, and plot the pairs of values on the coordinate plane. Use tables to compare ratios.	Lessons 4.3, 4.4, 4.5, 4.8
	b. Solve unit rate problems including those involving unit pricing and constant speed.	Lessons 4.6, 4.7
	c. Find a percent of a quantity as a rate per 100 (e.g., 30% of a quantity means 30/100 times the quantity); solve problems involving finding the whole, given a part and the percent.	Lessons 5.1, 5.2, 5.3, 5.4, 5.5, 5.6
	d. Use ratio reasoning to convert measurement units; manipulate and transform units appropriately when multiplying or dividing quantities.	Lessons 6.1, 6.2, 6.3, 6.4, 6.5

Standards You Will Learn

Student Edition Lessons

Domain: The Number System

Apply and extend previous understandings of multiplications and division to divide fractions by fractions.

6.NS.A.1	Interpret and compute quotients of fractions, and solve word problems involving division of fractions by fractions, e.g., by using visual fraction models and equations to represent the problem.	Lessons 2.5, 2.6, 2.7, 2.8, 2.9, 2.10

Compute fluently with multi-digit numbers and find common factors and multiples.

6.NS.B.2	Fluently divide multi-digit numbers using the standard algorithm.	Lesson 1.1
6.NS.B.3	Fluently add, subtract, multiply, and divide multi-digit decimals using the standard algorithm for each operation.	Lessons 1.6, 1.7, 1.8, 1.9
6.NS.B.4	Find the greatest common factor of two whole numbers less than or equal to 100 and the least common multiple of two whole numbers less than or equal to 12. Use the distributive property to express a sum of two whole numbers 1–100 with a common factor as a multiple of a sum of two whole numbers with no common factor.	Lessons 1.2, 1.3, 1.4, 1.5, 2.3, 2.4

Correlations H19

Standards You Will Learn

Student Edition Lessons

Apply and extend previous understandings of numbers to the system of rational numbers.

6.NS.C.5	Understand that positive and negative numbers are used together to describe quantities having opposite directions or values (e.g., temperature above/below zero, elevation above/below sea level, credits/ debits, positive/negative electric charge); use positive and negative numbers to represent quantities in real-world contexts, explaining the meaning of 0 in each situation.	Lesson 3.1, 3.3
6.NS.C.6	Understand a rational number as a point on the number line. Extend number line diagrams and coordinate axes familiar from previous grades to represent points on the line and in the plane with negative number coordinates.	
	a. Recognize opposite signs of numbers as indicating locations on opposite sides of 0 on the number line; recognize that the opposite of the opposite of a number is the number itself, e.g., −(−3) = 3, and that 0 is its own opposite.	Lessons 3.1, 3.3
	b. Understand signs of numbers in ordered pairs as indicating locations in quadrants of the coordinate plane; recognize that when two ordered pairs differ only by signs, the locations of the points are related by reflections across one or both axes.	Lesson 3.8
	c. Find and position integers and other rational numbers on a horizontal or vertical number line diagram; find and position pairs of integers and other rational numbers on a coordinate plane.	Lessons 2.1, 3.1, 3.3, 3.7

Standards You Will Learn

Student Edition Lessons

Apply and extend previous understandings of numbers to the system of rational numbers. *(Continued)*

6.NS.C.7	Understand ordering and absolute value of rational numbers.	
	a. Interpret statements of inequality as statements about the relative position of two numbers on a number line diagram.	Lessons 2.2, 3.2, 3.4
	b. Write, interpret, and explain statements of order for rational numbers in real-world contexts.	Lessons 3.2, 3.4
	c. Understand the absolute value of a rational number as its distance from 0 on the number line; interpret absolute value as magnitude for a positive or negative quantity in a real-world situation.	Lesson 3.5
	d. Distinguish comparisons of absolute value from statements about order.	Lesson 3.6
6.NS.C.8	Solve real-world and mathematical problems by graphing points in all four quadrants of the coordinate plane. Include use of coordinates and absolute value to find distances between points with the same first coordinate or the same second coordinate.	Lessons 3.9, 3.10

Correlations H21

Standards You Will Learn

Student Edition Lessons

Domain: Expressions and Equations

Apply and extend previous understandings of arithmetic to algebraic expressions.

6.EE.A.1	Write and evaluate numerical expressions involving whole-number exponents.	Lessons 7.1, 7.2
6.EE.A.2	Write, read, and evaluate expressions in which letters stand for numbers.	
	a. Write expressions that record operations with numbers and with letters standing for numbers.	Lesson 7.3
	b. Identify parts of an expression using mathematical terms (sum, term, product, factor, quotient, coefficient); view one or more parts of an expression as a single entity.	Lesson 7.4
	c. Evaluate expressions at specific values of their variables. Include expressions that arise from formulas used in real-world problems. Perform arithmetic operations, including those involving whole-number exponents, in the conventional order when there are no parentheses to specify a particular order (Order of Operations).	Lessons 7.5, 10.1, 10.3, 10.5, 10.6, 10.7, 11.3, 11.4, 11.6
6.EE.A.3	Apply the properties of operations to generate equivalent expressions.	Lessons 7.7, 7.8

Standards You Will Learn

Student Edition Lessons

Apply and extend previous understandings of arithmetic to algebraic expressions. *(Continued)*

6.EE.A.4	Identify when two expressions are equivalent (i.e., when the two expressions name the same number regardless of which value is substituted into them). *For example, the expressions y + y + y and 3y are equivalent because they name the same number regardless of which number y stands for.*	Lesson 7.9

Reason about and solve one-variable equations and inequalities.

6.EE.B.5	Understand solving an equation or inequality as a process of answering a question: which values from a specified set, if any, make the equation or inequality true? Use substitution to determine whether a given number in a specified set makes an equation or inequality true.	Lessons 8.1, 8.8
6.EE.B.6	Use variables to represent numbers and write expressions when solving a real-world or mathematical problem; understand that a variable can represent an unknown number, or, depending on the purpose at hand, any number in a specified set.	Lesson 7.6
6.EE.B.7	Solve real-world and mathematical problems by writing and solving equations of the form $x + p = q$ and $px = q$ for cases in which p, q and x are all nonnegative rational numbers.	Lessons 8.2, 8.3, 8.4, 8.5, 8.6, 8.7, 10.1
6.EE.B.8	Write an inequality of the form $x > c$ or $x < c$ to represent a constraint or condition in a real-world or mathematical problem. Recognize that inequalities of the form $x > c$ or $x < c$ have infinitely many solutions; represent solutions of such inequalities on number line diagrams.	Lessons 8.9, 8.10

Correlations H23

Standards You Will Learn

Student Edition Lessons

Represent and analyze quantitative relationships between dependent and independent variables.

6.EE.C.9	Use variables to represent two quantities in a real-world problem that change in relationship to one another; write an equation to express one quantity, thought of as the dependent variable, in terms of the other quantity, thought of as the independent variable. Analyze the relationship between the dependent and independent variables using graphs and tables, and relate these to the equation. *For example, in a problem involving motion at constant speed, list and graph ordered pairs of distances and times, and write the equation d = 65t to represent the relationship between distance and time.*	Lessons 9.1, 9.2, 9.3, 9.4, 9.5

Domain: Geometry

Solve real-world and mathematical problems involving area, surface area, and volume.

6.G.A.1	Find the area of right triangles, other triangles, special quadrilaterals, and polygons by composing into rectangles or decomposing into triangles and other shapes; apply these techniques in the context of solving real-world and mathematical problems.	Lessons 10.1, 10.2, 10.3, 10.4, 10.5, 10.6, 10.7, 10.8, 11.7
6.G.A.2	Find the volume of a right rectangular prism with fractional edge lengths by packing it with unit cubes of the appropriate unit fraction edge lengths, and show that the volume is the same as would be found by multiplying the edge lengths of the prism. Apply the formulas $V = lwh$ and $V = bh$ to find volumes of right rectangular prisms with fractional edge lengths in the context of solving real-world and mathematical problems.	Lessons 11.5, 11.6, 11.7

Standards You Will Learn

Student Edition Lessons

Solve real-world and mathematical problems involving area, surface area, and volume. (Continued)

6.G.A.3	Draw polygons in the coordinate plane given coordinates for the vertices; use coordinates to find the length of a side joining points with the same first coordinate or the same second coordinate. Apply these techniques in the context of solving real-world and mathematical problems.	Lesson 10.9
6.G.A.4	Represent three-dimensional figures using nets made up of rectangles and triangles, and use the nets to find the surface area of these figures. Apply these techniques in the context of solving real-world and mathematical problems.	Lessons 11.1, 11.2, 11.3, 11.4, 11.7

Domain: Statistics and Probability

Develop understanding of statistical variability.

6.SP.A.1	Recognize a statistical question as one that anticipates variability in the data related to the question and accounts for it in the answers. *For example, "How old am I?" is not a statistical question, but "How old are the students in my school?" is a statistical question because one anticipates variability in students' ages.*	Lesson 12.1
6.SP.A.2	Understand that a set of data collected to answer a statistical question has a distribution which can be described by its center, spread, and overall shape.	Lessons 12.6, 13.1, 13.4, 13.6, 13.7, 13.8
6.SP.A.3	Recognize that a measure of center for a numerical data set summarizes all of its values with a single number, while a measure of variation describes how its values vary with a single number.	Lessons 12.6, 13.4, 13.6

Standards You Will Learn

Student Edition Lessons

Summarize and describe distributions.

6.SP.B.4	Display numerical data in plots on a number line, including dot plots, histograms, and box plots.	Lessons 12.3, 12.4, 12.8, 13.2
6.SP.B.5	Summarize numerical data sets in relation to their context, such as by:	
	a. Reporting the number of observations.	Lesson 12.2
	b. Describing the nature of the attribute under investigation, including how it was measured and its units of measurement.	Lesson 12.2
	c. Giving quantitative measures of center (median and/or mean) and variability (interquartile range and/or mean absolute deviation), as well as describing any overall pattern and any striking deviations from the overall pattern with reference to the context in which the data were gathered.	Lessons 12.5, 12.6, 12.7, 13.1, 13.3, 13.4, 13.8
	d. Relating the choice of measures of center and variability to the shape of the data distribution and the context in which the data were gathered.	Lessons 12.7, 13.5, 13.7

Common Core State Standards © Copyright 2010. National Governors Association Center for Best Practices and Council of Chief State School Officers. All rights reserved. This product is not sponsored or endorsed by the Common Core State Standards Initiative of the National Governors Association Center for Best Practices and the Council of Chief State School Officers.

Index

A

Absolute value, 165–168
    compare, 171–174
    defined, 138, 165

Activity, 95, 108, 113, 171, 184, 211, 533, 565, 656, 745

Addition
    Addition Property of Equality, 420, 440
    Associative Property of, 401
    Commutative Property of, 401
    decimals, 37–40
    Distributive Property, 401–404
    equations
        model and solve, 433–436
        solution, 421–424
    Identity Property of, 401
    order of operations, 38–39, 363
    properties of, 401

Addition Properties, 355

Addition Property of Equality, 420, 440

Algebra
    algebraic expressions
        combine like terms, 395–398
        defined, 369
        equivalent
            generate, 401–404
            identifying, 407–410
        evaluating, 381–384, 419, 531
        exponents, 357–360
        identifying parts, 375–378
        like terms, 395–398
        simplifying, 395–398
        terms of, 376
        translating between tables and, 370, 378
        translating between words and, 369–372
        use variables to solve problems, 389–392
        variables in, 369, 389–392
        writing, 369–372
    area
        composite figures, 571–574
        defined, 532
        parallelograms, 533–536
        rectangles, 533–536
        regular polygons, 565–568
        squares, 533–536, 595
        surface area, 603–606, 609–612, 615–618
        trapezoids, 551–554, 557–560
        triangles, 539–542, 545–548, 595
    distance, rate, and time formulas, 341–344
    equations
        addition, 433–442
        defined, 420
        division, 451–454
        with fractions, 457–460
        multiplication, 445–448, 451–454
        solution of, 420, 421–424
        subtraction, 439–442
        write from word sentence, 428–429
        writing, 427–430
    equivalent ratios
        graph to represent, 255–258
        to solve problems, 229–232
    evaluate, 121
    integers
        absolute value, 165–168
        compare and order, 145–148
        defined, 138, 139, 163
        opposites, 138
        order of operations, 38, 39, 44, 45, 51, 57, 82, 83, 109, 121, 363, 364, 381, 705
    inverse operations, 7
    least common multiple, 19
    patterns
        divide mixing patterns, 121
    proportions, equivalent ratios to solve, 235–238
    reasoning, 191
        finding least common multiple, 19
        finding the missing number, 167
    surface area, 603–606, 609–612, 615–618
    unit rates to solve problems, 249–252
    volume, 629–632

Algebraic expressions
    combine like terms, 395–398
    defined, 369
    equivalent
        generate, 401–404
        identifying, 407–410
    evaluating, 381–384, 419, 531
    exponents, 357–360
    identifying parts, 375–378
    like terms, 395–398
    simplifying, 395–396
    terms of, 376
    translating between tables and, 370, 378
    translating between words and, 369–372
    use variables to solve problems, 389–392
    variables in, 369, 389–392
    writing, 369–372

Area
    composite figures, 571–574
    defined, 532
    of parallelograms, 533–536
    of rectangles, 595

Index H27

of regular polygons, 565–568
of squares, 533–536, 595
surface area, 603–606, 609–612, 615–618
of trapezoids, 551–554, 557–560
of triangles, 539–542, 545–548, 595

Art. Connect to, 284

Associative Property
of Addition, 401
of Multiplication, 401

Balance point
mean as, 675–678

Bar graph
reading, 647

Bar model, 250–251, 289, 295, 297

Base
of a number, 356, 357
of solid figure, 597–600

Basic fact, 11

Benchmark, 68, 82

Benchmark fractions, 67

Box plots, 713–716
defined, 706, 713
for finding interquartile range, 726–727, 734
for finding range, 726–727, 734

Bubble map, 68, 210, 268, 356, 490, 532, 596

Capacity
converting units of, 321–324
customary units of, 321
defined, 314
metric units of, 322

Cause and effect, 500

Centigrams, 328, 349

Centiliter, 322

Centimeters, 316

Chapter Review/Test, 61–66, 131–136, 201–206, 261–266, 307–312, 347–352, 413–418, 483–488, 523–528, 589–594, 641–646, 699–704, 757–762

Checkpoint, Mid-Chapter. *See* **Mid-Chapter Checkpoint**

Cluster, 707–710

Coefficient, 356, 376

Combine like terms, 395–398

Common Core State Standards, H17–H26

Common denominator, 75, 93, 138

Common factor, 4, 23

Commutative Property
addition, 355, 401
multiplication, 401, 419

Compare
absolute values, 171–174
decimals, 137
fractions, 137
integers, 145–148
and order fractions and decimals, 75
and order whole numbers, 67
rational numbers, 157–160
whole numbers, 67

Compatible numbers
defined, 55, 68, 101
estimate quotients, 101–104

Composite figures, 532
area of, 571–574
defined, 571

Concepts and Skills, 35–36, 93–94, 163–164, 241–242, 287–288, 333–334, 387–388, 463–464, 509–510, 563–564, 621–622, 673–674, 731–733

Congruent, 532, 539

Connect
to Art, 284
to Health, 84
to Reading, 174, 214, 338, 474, 500, 658
to Science, 40, 58, 360, 568, 632, 710

Conversion factor, 315

Conversions
capacity, 321–324
conversion factor, 314
length, 315–318
mass, 327–330
weight, 327–330

Coordinate Grid
identify points, 489

Coordinate plane, 138
defined, 177
diagram to solve problems on, 195–198
distance, 189–192
figures, 583–586
horizontal line, 189–192
linear equations, 517–520
ordered pair relationships, 183–186
problem solving, 195–198
rational numbers, 177–180
vertical line, 189–192

Correlations,
Common Core State Standards, H17–H26

Cross-Curricular Connections
Connect to Art, 284
Connect to Health, 84

Connect to Reading, 174, 214, 338, 474, 500, 658
Connect to Science, 40, 58, 360, 568, 632, 710

Cube, 596
    net, 599
    surface area, 604, 610
    volume, 630

Cup, 321

Customary units of measure
    capacity, 321–324
    converting, 315–330
    length, 315–318
    weight, 327–330

D

Data collection
    description, 655–658
    frequency table, 662–664, 668, 669, 670, 700, 702
    graphs
        histogram, 667–670
    mean, 648, 675–678, 681
    median, 648, 681
    mode, 648, 681
    range, 726
    relative frequency table, 662–664

Data sets
    box plots, 713–716
    cluster, 707–710
    collection, 655–658
    distributions, 745–754
    dot plots, 661–664
    frequency tables, 662–664
    gap, 707–710
    graphs
        histogram, 667–670
    interquartile range, 726–728
    mean, 648, 675–678, 681
    mean absolute deviation, 719–722
    measure of variability, 725–728
    measures of center, 681–684, 739–742
    median, 681
    outliers, 687–690
    patterns, 707–710
    peak, 708–710
    problem solving, 693–696
    range, 725–727
    relative frequency table, 662–664
    statistical question, 649–652

Decagon, 566

Decigrams, 328, 349

Deciliter, 322

Decimal Places
    counting, 44

Decimals, 4
    addition, 37–40
    compare, 647
        fractions and, 75–78
        rational numbers, 157–160
    converting to fractions, 69–72
    division
        estimate with compatible numbers, 55–58
        multi-digit numbers, 5–8
        by whole numbers, 49–52
    fractions, 75–78
    model, 267
    multiplication, 43–46
        Distributive Property, 24, 25
        estimate, 43–49
        whole numbers, 267
    order
        fractions, 75–78
        rational numbers, 157–160
    percent written as, 275–278
    place value, 37
    placing zeros, 38
    round, 3, 43
    subtraction, 37–40
    write as percent, 281–284

Decimeter, 316

Dekaliter, 322

Dekameter, 316

Denominator, 68

Dependent variable, 490, 491–494

Diagrams
    ladder diagram, 12
    tree, 420
    Venn diagrams, 17, 23, 26, 62

Distance
    coordinate plane, 189–191
    distance, rate, and time formulas, 341–344

Distribution
    data set, 745–748

Distributive Property
    addition, 24, 25, 31
    multiplication, 24, 402

Dividends, 4

Divisibility
    rules for, 11, 12

Divisible, 4

Division
    decimals
        estimate with compatible numbers, 55–58
        whole numbers, 49–52
    Division Property of Equality, 451
    equations, 451–454
    to find equivalent fractions, 209
    finding quotient, 647

fractions, 107–110
   estimate quotients with compatible numbers, 101–104
   mixed numbers, 68, 69, 113–116
   model, 94–98
   multiplicative inverse, 108
   reciprocal, 108
as inverse operation, 7
mixed numbers, 119–122
model mixed number, 113–116
multi-digit numbers, 5–8
order of operations, 363

Division Property of Equality, 451

Divisors, 4

Dot plots, 648
   defined, 661
   finding mean absolute deviation, 720–721
   finding outliers, 687–690

Double number line, 301

Draw Conclusions, 96, 114, 212, 214, 270, 434, 446, 540, 552, 603, 624, 675, 720

Equations
   addition, 439–442
   addition, models to solve, 433–436
   defined, 420
   division, 451–454
   with fractions, 457–460
   linear, 490
   multiplication, 445–448, 451–454
   to represent dependent and independent variable, 491–494
   solution, 420, 421–424
   subtraction, 439–442
   translate between graphs and, 517–520
   translating between tables and, 497–500
   writing, 427–430

Equivalent
   defined, 75
   expressions, 401–404, 407–410
   fractions, 137, 138

Equivalent algebraic expressions
   defined, 401
   identifying, 407–410
   writing, 401–404

Equivalent fractions, 68, 93, 137, 210
   defined, 75, 268
   divide to find, 209
   multiply to find, 209

Equivalent ratios, 223–226, 255–258
   defined, 210
   finding
      by multiplication table, 223–224
      by unit rate, 243–246
   graph to represent, 255–258
   to solve problems, 229–232
   use, 235–238

Error Alert, 56, 88, 189, 224, 282, 316, 382, 428, 492, 603, 713, 740

Estimation, 4
   decimals
      addition and subtraction, 37–40
      division, 49–52
      multiplication, 43–46
   fractions
      division, 107–110
      multiplication, 81–84
   quotients, 101–104
   using compatible numbers, 5–6, 101–104

Evaluate, 356, 363

Exponents, 357–360
   defined, 357
   order of operations, 363
   write repeated multiplication, 357–360

Expressions
   algebraic, 369–372
   equivalent, 401–404, 407–410
   evaluate, 381–384, 489, 595
   identifying parts, 375–378
   numerical, 363–366
   terms, 376
   writing, 369–372

Factors
   common, 4, 23
   defined, 23
   greatest common factor, 23–26
   least common multiple, 4
   prime, 12
   prime factorization, 11–14, 17–20
   simplify, 87–90

Factorization, prime. See Prime factorization

Factor tree, 11–14, 61

Fair share
   mean, 675–678

Feet, 315

Flow map, 4, 138

Fluid ounces, 321

H30 Index

Foot. *See* **Feet**
Formula
    distance, 341–344
    rate, 341–344
    time, 341–344
Fractions
    compare, 137
        decimals and, 75–78
        rational numbers, 157–160
    converting to decimals, 69–72
    denominator of, 69
    division, 107–110
        mixed number, 113–116, 119–122
        model, 95–98, 113–116
        multiplicative inverses, 108–110
        reciprocals, 108–110
    equations with, 457–460
    equivalent, 68, 93, 137
    mixed numbers
        converting to decimals, 69–72
        defined, 69
        division, 113–116, 119–122
        multiplication, 81–83
    model fraction division, 95–98
    multiplication, 67
        estimate products, 81–84
        simplifying before, 87–90
        whole numbers, 67
    operations, 125–128
    order
        decimals, 75–78
    percent written as, 275–278
    problem solving, fraction operations, 125–128
    product of two, 81
    rates, 218–220
    ratios written as, 217
    in simplest form, 69, 71
    unlike, 75–76
    volume, 623–626
    write as percent, 281–284
    writing, 69
    writing as decimal, 70
Frequency, 661
Frequency tables, 661–664
Functions
    cause and effect, 500
    graphing, 511–514
    linear equations, 517–520

Gallon, 314
Gap, 707–710
GCF. *See* **Greatest common factor (GCF)**

Generalization, Make, 474
Geometric measurements, 635–638
Geometry
    area
        composite figures, 571–574
        defined, 533
        parallelogram, 533–536
        rectangles, 533–536
        regular polygon, 565–568
        squares, 533–536
        trapezoids, 551–554, 557–560
        triangles, 539–542, 545–548
    composite figures
        area, 571–574
    figures on coordinate plane, 583–586
    solid figures
        defined, 597
        nets, 597–600
        pyramid, 598–600
        rectangular pyramid, 598–600
        triangular prism, 597–600
        volume, 630
    surface area, 603–606, 615–618
    volume
        defined, 623
        prism, 629–632
        rectangular prisms, 623–626, 629–632
Go Deeper, In some Student Edition lessons. Some examples are: 19, 360, 696
Grams
    as metric unit, 328
    solving problems, 328–329
Graphic organizer
    Bubble Map, 68, 210, 268, 356, 490, 532, 596
    Chart, 648, 706
    Flow map, 4, 138
    Tree diagram, 420
    Venn diagram, 17, 23, 26, 62, 314
Graph relationships, 511–514
Graphs
    bar, 647
    box plots, 713–716
    equations and, 517–520
    equivalent ratios, 255–258
    histogram, 648, 667–670, 708
    inequalities, 477–480
    linear equations, 517–520
    points on coordinate plane, 183–186
    relationships, 511–514
    to represent equivalent ratios, 255–258
Greatest common factor (GCF), 4, 23–26, 35, 61
    defined, 23
    to express sum as a product, 25
    problem solving, 29–32

Health. Connect to, 84
Hectograms, 328, 349
Hectoliter, 322
Hectometer, 316
Hexagon, 531
Histogram, 648, 667–670, 708
Horizontal line
    coordinate plane, 189–192

Identity Property
    Addition, 401
    Multiplication, 401, 419
Inches, 315
Independent variable, 490, 491–494
Inequalities, 420
    defined, 465
    graphing, 477–480
    solutions, 465–468
    writing, 471–473
Input-output table, 497–499, 523
Input value, 497
Integers
    absolute value, 165–168
        compare, 171–173
    compare and order, 145–148
    defined, 138, 139, 163
    negative, 139
    opposites, 138
    order of operations, 363, 364, 381, 705
    positive, 139
Interquartile range, 706, 726
Inverse operations, 7, 420
    fraction division, 108
Investigate, 95, 113, 211, 269, 433, 445, 539, 551, 603, 623, 675, 719
Isosceles triangle, 184

Kilograms, 328, 329, 333
Kiloliter, 322
Kilometer, 316

Ladder diagram, 12
Lateral area
    of triangular pyramid, 616
Least common multiple (LCM), 4, 35, 61
    defined, 17
    finding, 18, 19
    using a list, 17
    using prime factorization, 17
    using Venn diagram, 17
Length
    converting units, 313, 315–318
    customary units, 315
    metric units, 316, 329
Like terms
    combining, 395–398
    defined, 395
Linear equation
    defined, 490, 517
    graphing, 517–520
Line of symmetry, 184
Line plot. *See* **Dot plots**
Line symmetry, 184
Liter, 314
Lower quartile, 713

Make Connections, 96, 114, 212, 270, 434, 446, 540, 552, 604, 624, 676, 720
Mass
    converting units, 327–333
    defined, 314
    metric units, 328
Materials
    algebra tiles, 433, 445
    centimeter grid paper, 603
    counters, 211, 675, 719
    cubes, 623
    fraction strips, 95
    grid paper, 533, 551
    large number line from 0–10, 719
    MathBoard, 433, 445
    net of rectangular prisms, 623
    pattern blocks, 113
    ruler, 539, 551, 603, 656
    scissors, 533, 539, 551, 603, 623
    tape, 623
    tracing paper, 539
    two-color counters, 211

MathBoard. In every student edition. Some examples are: 6, 7, 12, 71, 77, 83, 141, 147, 153, 212, 218, 225, 271, 276, 283, 317, 323, 329, 359, 365, 371, 429, 435, 441, 493, 499, 505, 535, 541, 547, 599, 605, 611, 651, 657, 663, 709, 715, 721

Mathematical Practices
1. Make sense of problems and persevere in solving them. In many lessons. Some examples are: 6, 104, 121, 558, 600
2. Reason abstractly and quantitatively. In many lessons. Some examples are: 19, 632
3. Construct viable arguments and critique the reasoning of others. In many lessons. Some examples are: 88, 166, 498, 747
4. Model with mathematics. In many lessons. Some examples are: 24, 96, 396
5. Use appropriate tools strategically. In many lessons. Some examples are: 251, 330
6. Attend to precision. In many lessons. Some examples are: 45, 243, 611
7. Look for and make use of structure. In many lessons. Some examples are: 13, 256
8. Look for and express regularity in repeated reasoning. In many lessons. Some examples are: 109, 192, 359, 540, 552, 707

Math Idea, 5, 18, 76, 139, 145, 165, 315, 336, 357, 369, 402, 421, 465, 533, 584, 661, 733

Math in the Real World Activities, 3, 67, 137, 209, 267, 313, 355, 419, 489, 531, 595, 647, 705

Math on the Spot videos. In every student edition lesson. Some examples are: 8, 14, 20, 72, 78, 84, 142, 148, 154, 214, 220, 226, 272, 278, 284, 318, 324, 338, 360, 366, 372, 424, 430, 442, 494, 497, 506, 536, 542, 554, 600, 606, 612, 652, 658, 664, 710, 716, 722

Math Talk. In all Student Edition lessons, 7, 11, 12, 13, 17, 19, 23, 25, 29, 38, 39, 44, 49, 51, 56, 57, 69, 70, 71, 75, 76, 77, 81, 83, 87, 89, 96, 103, 107, 108, 109, 113, 114, 120, 121, 125, 139, 140, 141, 145, 146, 147, 151, 152, 153, 157, 158, 159, 165, 167, 172, 173, 178, 179, 184, 185, 189, 190, 191, 195, 196, 211, 212, 218, 219, 223, 225, 229, 236, 237, 243, 245, 249, 255, 257, 270, 276, 282, 289, 290, 295, 296, 301, 302, 303, 315, 316, 317, 322, 323, 328, 329, 336, 341, 342, 357, 358, 359, 363, 365, 369, 371, 376, 377, 381, 382, 383, 389, 390, 391, 395, 402, 403, 407, 408, 409, 421, 422, 423, 427, 429, 433, 434, 439, 441, 445, 446, 451, 452, 453, 457, 458, 460, 465, 466, 467, 471, 472, 478, 479, 491, 493, 497, 503, 504, 511, 512, 513, 517, 519, 533, 535, 540, 545, 557, 558, 559, 565, 567, 571, 573, 577, 578, 599, 604, 609, 611, 615, 617, 623, 624, 629, 631, 635, 636, 650, 651, 656, 657, 662, 669, 675, 681, 683, 707, 714, 715, 720, 725, 726, 727, 734, 735, 739, 745, 746, 747, 751

Mean
defined, 648, 681
as fair share and balance point, 675–678
finding, 681–684
set of data, 705

Mean absolute deviation, 719–722
defined, 720
dot plot, 721–722

Measurement
conversion factor, 314
converting units of capacity, 321–323
converting units of length, 313, 315–318
converting units of mass, 327–330
converting units of volume, 313
converting units of weight, 313, 327–333

Measure of center, 681–684
applying, 739–742
defined, 681
effect of outliers, 687–690

Measure of variability, 706, 725–728
applying, 739–742
choose an appropriate, 733–736
defined, 725

Median
defined, 648, 681
finding, 681–684
outlier, 687–690

Meter, 314, 322

Metric units of measure
capacity, 322
converting, 316, 322
length, 316–317
mass, 328

Mid-Chapter Checkpoint, 35–36, 93–94, 163–164, 241–242, 287–288, 333–334, 387–388, 463–464, 509–510, 563–564, 621–622, 673–674, 731–732

Miles, 315

Mililiters, 322

Milligrams, 328

Millimeters, 316

Mixed numbers, 68
converting to decimals, 69–72
division, 119–122
model division, 113–116
writing, 69

Mode
defined, 648, 681
finding, 681–684

Model fraction division, 95–98

Model mixed number division, 113–116

Model percents, 269–272

Model ratios, 211–214

Multi-digit decimals
adding and subtracting, 37–40
multiplication, 43–46

Multi-Digit Numbers
division, 5–8

Multiplication
Associative Property, 401, 404
Commutative Property, 401, 407, 419
decimals, estimate, 43–49
Distributive Property, 402–404, 408–410
equations, 445–452
  model and solve, 445–448
  solution, 451–454
exponents as repeated, 357–360
fractions, products, 81–84
fractions and whole numbers, 67
Identity Property of, 401, 419
inverse operation, 7
order of operations, 363
prime factorization, 12
Properties of Multiplication, 401, 419, 452
simplify, 87–90
table to find equivalent ratios, 223–226

Multiplication tables, 223–226

Multiplicative inverse, 68, 108

Negative numbers, 138

Nets, 597–600
defined, 596
surface area, 603–606
surface area of cube, 610
surface area of prism, 605
rectangular pyramid, 598
triangular prism, 597–600

Number line, 151–154
absolute value, 165–168
compare and order
  fractions and decimals, 75–78
  integers, 145–148
  rational numbers, 157–160
divide by using, 107–110
find quotient, 109
inequalities, 477–480
negative integers, 139–142
positive integers, 139–142
rational numbers, 151–154

Number Patterns, 489

Numbers
compatible, 4, 5, 6
negative, 138, 139–142
positive, 139–142

Numerators, 68, 75, 81, 87–90

Numerical expression
defined, 356, 363
order of operations, 363–366
simplifying, 363–366

On Your Own, In every Student Edition lesson. Some examples are: 7, 13, 19, 71, 77, 83, 141, 147, 153, 219, 225, 232, 277, 283, 291, 317, 323, 329, 359, 365, 371, 423, 429, 441, 493, 499, 506, 535, 547, 559, 599, 611, 617, 651, 657, 663, 709, 715, 727

Opposites
defined, 138, 139
negative integers, 139
positive integers, 139

Order
fractions and decimals, 75–78
integers, 145–148
rational numbers, 157–160

Ordered pairs, 138, 177, 210
relationships, 183–186

Order of operations
algebraic expressions, 381–384
integers, 363, 364, 381, 705
numerical expressions, 363–366

Origin, 177

Ounces, 314, 321, 327–330

Outliers, 648
defined, 687
effect, 687–690

Output value, 497

Parallelogram
area, 533–536
defined, 532

Parentheses
order of operations, 355

Pattern blocks, 113, 115

Patterns
changing dimensions, 577–580
data, 707–710
extend, 209
finding, 108
number, 489
use, 72

H34 Index

Peak, 707–710

Pentagon, 531

Percents
  bar model, 295–297
  defined, 268, 269
  find the whole, 301–304
  model, 269–272
  of a quantity, 289–292
  solve problem with model, 295–298
  write as decimal, 275–278
  write as fraction, 275–278
  write decimal as, 281–284
  write fraction as, 281–284

Perimeter
  finding, 531

Personal Math Trainer, In all Student Edition chapters. Some examples are: 3, 344, 742

Pint, 314, 321

Place value
  decimal, 69
  of a digit, 37

Polygons
  area, 565–568
  changing dimensions, 577–580
  coordinate plane, 583–586
  graphs on coordinate plane, 583–586
  identify, 531

Polyhedron, 596

Pose a Problem, 52, 97, 110, 226, 252, 272, 448, 506, 670, 716

Positive numbers, 139–142

Pound, 327, 329, 330

Practice and Homework
  In every Student Edition lesson. Some examples are: 9, 10, 73, 74, 143, 144, 215, 216, 273, 274, 319, 320, 361, 362, 425, 426, 495, 496, 537, 538, 601, 602, 653, 654, 712, 713

Preview words, 4, 68, 138, 210, 268, 314, 356, 420, 490, 532, 596, 648, 706

Prime factorization, 11–14, 17–20, 35
  defined, 4, 11
  divisibility rule, 11
  factor tree, 11
  finding, 12–13
  greater common factor, 23
  ladder diagram, 12
  least common multiple, 17–18
  reason for using, 18

Prime number, 4
  defined, 11

Prism, 596
  net, 597–600, 603–606
  surface area, 609–612
  volume, 629–632

Problem Solving
  analyze relationships, 503–506
  apply greatest common factor, 29–32
  changing dimensions, 577–580
  combine like terms, 395–398
  coordinate plane, 195–198
  data displays, 693–696
  distance, rate, and time formulas, 341–344
  equations with fractions, 457–460
  fraction operations, 125–128
  geometric measurements, 635–638
  misleading statistics, 751–754
  percents, 295–298
  use tables to compare ratios, 229–232

Problem Solving. Applications. In most lessons. Some examples are: 8, 14, 26, 70, 78, 104, 142, 148, 154, 220, 226, 271, 272, 278, 318, 330, 366, 378, 384, 392, 424, 430, 436, 514, 520, 536, 541, 553–554, 600, 605, 618, 652, 670, 677–678, 716, 722, 742, 748

Projects, 2, 208, 354, 530

Properties
  Associative Property of Addition, 355
  Associative Property of Multiplication, 401
  Commutative Property of Addition, 355
  Commutative Property of Multiplication, 401, 419
  Distributive Property, 24, 402
  Division Property of Equality, 451
  Identity Property, 401, 419
  Multiplication Property of Equality, 452
  Subtraction Property of Equality, 420

Pyramid, 596
  defined, 598
  surface area, 615–618

Quadrants, 138, 183–186

Quadrilateral, 532

Quart, 321

Quartile
  lower, 713
  upper, 706, 713

Quotients, 267
  compatible numbers to estimate, 101–104

Range
defined, 706, 726
interquartile, 726

Rates, 210, 217–220
defined, 218, 268
distance, rate, and time formulas, 341–344
unit rate, 217–220, 243–246, 249–252
writing, 217–220

Rational numbers
absolute value, 165–168
compare and order, 157–164, 168, 171–176, 200–202, 205, 206, 208
coordinate plane, 177–180
defined, 138, 163
number line, 151–154

Ratios, 210, 217–220
defined, 210, 268
equivalent
 defined, 210
 finding, 235–238
 graph to represent, 255–258
 use, 235–238
 using multiplication tables to find, 223–226
model, 211–214
percent as, 275
rates, 217–220
tables to compare, 229–232
writing, 217–220

Reading. Connect to, 174, 214, 338, 474, 500, 658

Read the Problem, 29, 30, 125, 126, 195, 196, 229, 230, 295, 296, 341, 342, 395, 396, 457, 458, 503, 504, 577, 578, 635, 636, 751

Real World. *See* **Connect,** to Science; **Problem Solving; Problem Solving. Applications; Unlock the Problem**

Reciprocals, 68, 108

Rectangles, 532
area, 533–535, 595

Rectangular prisms
surface area, 603–606
volume, 623–626, 629–631

Rectangular pyramid, 598

Regular polygon
area, 565–568
defined, 532, 565
in nature, 568

Relationships
analyze, 503–506
graph, 511–514
ordered pair, 183–186

Relative frequency table, 661

Remember, 11, 38, 81, 82, 512, 624

Review and Test. *See* **Chapter Review/Test; Mid-Chapter Checkpoint**

Review Words, 4, 68, 138, 210, 268, 314, 356, 420, 490, 532, 596, 648, 706

Round decimals, 3, 43

Science. Connect to, 40, 58, 360, 568, 632, 710

Sense or Nonsense?, 116, 142, 258, 436, 468, 542, 684, 710, 742

Set of data, 649

Share and Show, 6, 7, 12, 71, 77, 83, 141, 147, 153, 213, 219, 225, 271, 276, 283, 317, 323, 329, 359, 365, 371, 423, 429, 435, 479, 493, 499, 505, 513, 535, 541, 547, 599, 605, 611, 647, 651, 657, 709, 715, 721

Show What You Know, 3, 67, 137, 209, 267, 313, 355, 419, 489, 531, 595, 647, 705

Simplest form, 68, 69–72, 81–84, 87–90, 109

Simplifying
algebraic expressions, 395–396
fractions, 68, 69–72, 81–84, 87–90, 108–109, 268
numerical expressions, 363–366
order of operations, 363

Solid figures
defined, 597
nets, 597–600
pyramid, 598
rectangular prism, 597, 629–631
rectangular pyramid, 598
surface area, 603–606, 609–612, 615–618, 635–638
triangular prism, 597
volume, 623–626, 629–632, 635–638

Solution of equations, 421–424

Solutions of inequalities, 465–468

Solve the Problem, 29, 30, 52, 125, 126, 195, 196, 229, 230, 252, 272, 295, 296, 341, 342, 395, 396, 457, 458, 503, 504, 578, 635, 636, 751

Squares, 532
area, 533–536, 595

Statistical question
defined, 649
recognizing, 649–652

Student Help
Error Alert, 56, 88, 189, 224, 316, 382, 428, 492, 603, 713, 740
Math Idea, 5, 18, 76, 139, 145, 165, 302, 315, 336, 357, 369, 402, 421, 465, 533, 584, 661, 733

H36 Index

Math Talk, In every lesson. Some examples are: 7, 11, 71, 77, 140, 245, 377, 389, 715
Remember, 11, 23, 38, 81, 82, 512, 624, 708
Write Math, 14, 26, 32, 392, 460, 560, 579

Subtraction
decimals, 37–40
equations
model and solve, 439–442
solution, 421–424
order of operations, 363
solve addition and subtraction equations, 439–442

Subtraction Property of Equality, 420, 439

Summarize, 658

Surface area, 603–606
cubes, 610
defined, 603
net, 603–606
prisms, 605, 609–612
pyramids, 615–632
rectangular prism, 603, 605, 609
triangular prism, 610

Symmetry
and data, 708
line, 184

Tables
translating between equation, 497–500

Technology and Digital Resources,
iTools, 435, 721
See Math on the Spot

Terms
algebraic expressions, 376
defined, 356, 376
like terms, 395–398

Test and Review. See Chapter Review/Test

ThinkSmarter, In every Student Edition lesson. Some examples are: 8, 370, 753

ThinkSmarter+, In all Student Edition chapters. Some examples are: 14, 436, 742

Thousandth, 37, 38

Three-dimensional figures, 597–600

Time
distance, rate, and time formulas, 341–344

Ton, 327

Transform units, 335–338

Trapezoids
area, 532, 551–554, 557–560
defined, 551

Tree diagram, 420

Triangles
area, 539–542, 545–548, 595
congruent, 532, 539

Triangular prism
net, 597–600
surface area, 610

Try Another Problem, 30, 126, 196, 230, 296, 342, 396, 458, 504, 578, 636, 694, 751

Try This!, 18, 43, 88, 119, 140, 145, 218, 244, 250, 357, 363, 369, 427, 471, 477, 650, 681

Understand Vocabulary, 4, 68, 138, 210, 268, 314, 356, 420, 490, 532, 596, 648, 706

Unit rate
defined, 218
finding, 243–246
graph to represent, 255–258
solve problems, 249–252

Units of capacity, 321

Unlock the Problem, 5, 11, 17, 20–23, 37, 69, 81, 122, 139, 145, 151, 217, 223, 229, 275, 281, 315, 321, 324, 357, 363, 369, 421, 439, 442, 491, 494, 497, 533, 545, 548, 597, 609, 612, 649, 655, 661, 707, 713, 725

Upper quartile, 706, 713–716

Variables, 356, 369
and algebraic expressions, 369–372, 389–392
defined, 369
dependent, 490, 491–494
independent, 490, 491–494
solve problems, 389–392

Venn diagram, 17, 23, 26, 62, 314

Vertical line, 189–191

Visualize It, 4, 68, 138, 210, 268, 314, 356, 420, 490, 532, 596, 648, 706

Vocabulary
Chapter Vocabulary Cards, At the beginning of every chapter.
Vocabulary Builder, 4, 68, 138, 210, 268, 314, 356, 420, 490, 532, 596, 648, 706
Vocabulary Game, 4A, 68A, 138A, 210A, 268A, 314A, 356A, 420A, 490A, 532A, 596A, 648A, 706A

Volume
cube, 630
defined, 596, 623
fractions and, 623–626
prism, 630
rectangular prisms, 623–626, 629–631

Weight
converting units, 327–330
customary units, 327
defined, 314
units, 327–330

What If, 31, 43, 78, 96, 127, 342, 397, 459, 492, 505

What's the Error, 72, 104, 148, 154, 318, 366, 384, 424, 430, 454, 554, 606

Whole numbers, 138
compare, 67
dividing decimals by, 49–52
greatest common factor, 23–26
least common multiple, 17–20
multiplication
by decimals, 267

Word sentence
writing equation, 428–429
writing inequality, 471–474

Write Math, In every Student Edition lesson. Some examples are: 8, 14, 26, 78, 97, 98, 160, 220, 232, 297, 330, 343, 366, 378, 384, 430, 579, 580, 600, 625, 721

Writing
algebraic expressions, 369–372
equations, 427–430
equivalent algebraic equations, 401–404
inequalities, 471–473
ratios, and rates, 217–220

x-axis, 177
x-coordinate, 177

Yard, 315
y-axis, 177
y-coordinate, 177

H38 Index

Table of Measures

METRIC

Length

1 meter (m) = 1,000 millimeters (mm)
1 meter = 100 centimeters (cm)
1 meter = 10 decimeters (dm)
1 dekameter (dam) = 10 meters
1 hectometer (hm) = 100 meters
1 kilometer (km) = 1,000 meters

Capacity

1 liter (L) = 1,000 milliliters (mL)
1 liter = 100 centiliters (cL)
1 liter = 10 deciliters (dL)
1 dekaliter (daL) = 10 liters
1 hectoliter (hL) = 100 liters
1 kiloliter (kL) = 1,000 liters

Mass/Weight

1 gram (g) = 1,000 milligrams (mg)
1 gram = 100 centigrams (cg)
1 gram = 10 decigrams (dg)
1 dekagram (dag) = 10 grams
1 hectogram (hg) = 100 grams
1 kilogram (kg) = 1,000 grams

CUSTOMARY

Length

1 foot (ft) = 12 inches (in.)
1 yard (yd) = 3 feet
1 yard = 36 inches
1 mile (mi) = 1,760 yards
1 mile = 5,280 feet

Capacity

1 cup (c) = 8 fluid ounces (fl oz)
1 pint (pt) = 2 cups
1 quart (qt) = 2 pints
1 quart = 4 cups
1 gallon (gal) = 4 quarts

Mass/Weight

1 pound (lb) = 16 ounces (oz)
1 ton (T) = 2,000 pounds

TIME

1 minute (min) = 60 seconds (sec)
1 hour (hr) = 60 minutes
1 day = 24 hours
1 week (wk) = 7 days

1 year (yr) = about 52 weeks
1 year = 12 months (mo)
1 year = 365 days
1 decade = 10 years
1 century = 100 years
1 millennium = 1,000 years

SYMBOLS

$=$	is equal to	10^2	ten squared
$\neq$	is not equal to	10^3	ten cubed
$\approx$	is approximately equal to	2^4	the fourth power of 2
$>$	is greater than	$|{-4}|$	the absolute value of $^-4$
$<$	is less than	%	percent
$\geq$	is greater than or equal to	(2, 3)	ordered pair (x, y)
$\leq$	is less than or equal to	°	degree

FORMULAS

Perimeter and Circumference		Area	
Polygon	$P =$ sum of the lengths of sides	Rectangle	$A = lw$
		Parallelogram	$A = bh$
Rectangle	$P = 2l + 2w$	Triangle	$A = \frac{1}{2}bh$
Square	$P = 4s$	Trapezoid	$A = \frac{1}{2}(b_1 + b_2)h$
		Square	$A = s^2$

Volume		Surface Area	
Rectangular Prism	$V = lwh$ or Bh	Cube	$S = 6s^2$
Cube	$V = s^3$		